YOUR KNOWLEDGE HAS VALUE

- We will publish your bachelor's and
  master's thesis, essays and papers

- Your own eBook and book -
  sold worldwide in all relevant shops

- Earn money with each sale

Upload your text at www.GRIN.com
and publish for free

Shrey Naik, Varun Pratapsinha Jadhav

Regulation Of Grid Voltage using an integrated Wind-PV system as STATCOM in Distributed Generation systems.

A Novel Approach to Improving Power Quality in DG systems.

GRIN Publishing

Bibliographic information published by the German National Library:

The German National Library lists this publication in the National Bibliography; detailed bibliographic data are available on the Internet at http://dnb.dnb.de .

This book is copyright material and must not be copied, reproduced, transferred, distributed, leased, licensed or publicly performed or used in any way except as specifically permitted in writing by the publishers, as allowed under the terms and conditions under which it was purchased or as strictly permitted by applicable copyright law. Any unauthorized distribution or use of this text may be a direct infringement of the author s and publisher s rights and those responsible may be liable in law accordingly.

Imprint:

Copyright © 2014 GRIN Verlag GmbH
Print and binding: Books on Demand GmbH, Norderstedt Germany
ISBN: 978-3-656-83476-2

This book at GRIN:

http://www.grin.com/en/e-book/283560/regulation-of-grid-voltage-using-an-inte-grated-wind-pv-system-as-statcom

GRIN - Your knowledge has value

Since its foundation in 1998, GRIN has specialized in publishing academic texts by students, college teachers and other academics as e-book and printed book. The website www.grin.com is an ideal platform for presenting term papers, final papers, scientific essays, dissertations and specialist books.

Visit us on the internet:

http://www.grin.com/

http://www.facebook.com/grincom

http://www.twitter.com/grin_com

Regulation of Grid Voltage using an integrated Wind-PV system Based STATCOM in Distributed Generation System

Shrey Shrikant Naik

Department of EEE,Goa College of Engineering,Farmagudi,Goa

Varun Pratapsinha Jadhav

Department of EEE,Goa College of Engineering,Farmagudi,Goa

Abstract— In the proposed paper, it is described how a Solar PV Farm along with a battery storage system can be used to regulate grid voltage in a PV-Wind integrated distributed generation System.

At night time Solar PV system is normally dormant (i.e. it does not generate power) but the stored power in batteries can be utilised efficiently to regulate the common coupling voltage by means of a FACTS based Static Synchronous Compensator (STATCOM) thereby improving the power quality.

With advancements in RES and increasing DG systems to provide for load demand the quality of power has to be maintained to optimum value and this paper focuses purely on improving regulation of voltage without using external regulation devices but the installed RES system.

In order to implement and validate the concept of the prescribed paper, SIMULINK Tool has been used.

I. Introduction

Transmission grids worldwide are presently facing challenges in integrating such large scale renewable systems (Wind Farms (WF) and Solar Farms (SF)) due to their limited power transmission capacity. To increase the available power transfer limits/capacity of existing transmission line, series compensation and various Flexible AC Transmission System (FACTS) devices are being proposed.Utilities are presently facing a major challenge of grid integrating an increasing number of renewable-energy-based distributed generators (DGs) while ensuring stability, voltage regulation, and power quality. During the night time, feeder loads are usually much lower compared to daytime, while the wind farms (WFs) produce more power due to increased wind speeds. This potentially causes reverse power to flow from the point of common coupling (PCC) toward the main grid resulting in feeder voltages to rise above allowable limits. To allow further DG connections, utilities need to install expensive voltage regulating devices. Voltage-source inverters are essential components of PV solar farms (SFs), which provide solar power conversion during daytime (normal operation). However, PV SFs are practically inactive during night time and do not produce any real power output. The proposed concept is to use the existing SF inverter as a STATCOM during night time to regulate voltage variations at the PCC due to increased and intermittent WF power and/or by load variations. With the development of distributed generation systems, the renewable electricity from PV sources became a resource of energy in great demand. The current control scheme is mainly used in PV inverter applications for real power and reactive power control schemes. The emergence of windgeneration is the leading source of renewable energy in the power industry, Wind farms totalling hundreds, even thousands, of MW are now being considered. Double Fed Induction Generator is the main type ofwind generation currently in use (the other is conventional induction generators) due to their variable speed operation, four-quadrant active and reactive power capability, low-converter cost, and reduced power losses.

II. DG System Overview

Fig 1 and 2 shows a single line diagram of the integrated wind energy and PV system, both with battery storage scheme. The wind farm is modelled as a fully controlled converter-inverter based doubly fed induction generator and PV SF modelled as a Voltage fed Inverter.

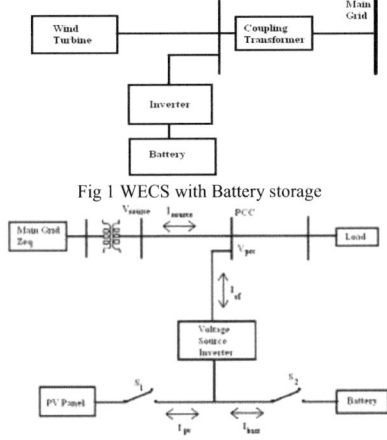

Fig 1 WECS with Battery storage

Fig 2 Operational Modes of the Solar Farm

III. SOLAR PV FARM AS A BATTERY CHARGER

A typical PV solar farm is basically inactive during night time and the bidirectional inverter used to deliver the PV DC power as three-phase AC power to the grid remains unutilized as well. Fig. 2 shows thepossible operational modes of the solar farm. The point at which the solar farm is connected to the grid is called the point of common coupling (PCC). In Fig.2, V_S and I_S represents the voltage and current at the secondary of the distribution transformer; V_{PCC} and V_L denote voltages at PCC and load terminal respectively; and I_{PV} is the current delivered by the PV solar panels ac current drawn/delivered by the solar farm inverter and the DC current flowing through the storage battery are represented by I_{SF} and I_{Batt}, respectively. Here a storage battery is connected on DC side of the solar farm inverter. Switch S_1 inFig.2is utilized to disconnect the PV solar panels especially during night-time and to charge the storage batteries from the main grid. [2]

IV. DOUBLE FED INDUCTION GENERATOR

The Double Fed Induction Generator (DFIG) is a generating principle widely used in wind turbines. It is based on an induction generator with a multiphase wound rotor and a multiphase slip ring assembly with brushes for access tothe rotor windings. It is possible to avoid the multiphase slip ring assembly but there are problems with efficiency, cost and size. A better alternative is a brushless wound-rotor doubly-fed electric machine.The principle of the DFIG is that the rotor windings are connected to the grid via slip rings and the back-to-back voltage source converter that controls both the rotor and the grid currents. Thus rotor frequency can freely differ from the grid frequency (50 or 60 Hz). By using the converter to control the rotor currents, it is possible to adjust the active and reactive power fed to the grid from the stator independently of the generator's turning speed. The control principle used is either the two-axis current vector control or direct torque control (DTC). DTC has turned out to have better stability than current vector control, especially when high reactive currents are required from the generator.The doubly-fed generator rotors are typically wound with 2 to 3 times the number of turns of the stator. This means that the rotor voltages will be higher and currents respectively lower. Thus in the typical ± 30 % operational speed range around the synchronous speed, the rated current of the converter is accordingly lower which leads to a lower cost of the converter. The drawback is that controlled operation outside the operational speed range is impossible because of the higher than rated rotor voltage. Further, the voltage transients due to the grid disturbances (three-phase and two-phase voltage dips, especially) will also be magnified. In order to prevent high rotor voltages and high currents resulting from these voltages from destroying the IGBTs and diodes of the converter, a protection circuit (called crowbar) is used. The crowbar will short-circuit the rotor windings through a small resistance when excessive currents or voltages are detected. In order to be able to continue the operation as quickly as possible an active crowbar has to be used. The active crowbar can remove the rotor short in a controlled way and thus the rotor side

converter can be started only after 20-60 ms from the start of the grid disturbance. Thus it is possible to generate reactive current to the grid during the rest of the voltage dip and in this way help the grid to recover from the fault. The AC/DC/AC converter is divided into two components: the rotor-side converter and the grid-side converter. The Voltage-Sourced Converters that use forced-commutated power electronic devices (IGBTs) to synthesize an AC voltage from a DC voltage source. A capacitor connected on the DC side acts as the DC voltage source. A coupling inductor $L_F$is used to connect grid side converter tothe grid. The three-phase rotor winding is connected to rotor side converter by slip rings and brushes and the three-phase stator winding isdirectly connected to the grid. The power captured by the wind turbine is converted into electrical power by the induction generator and it is transmitted to the grid by the stator and the rotor windings.

V. STATCOM

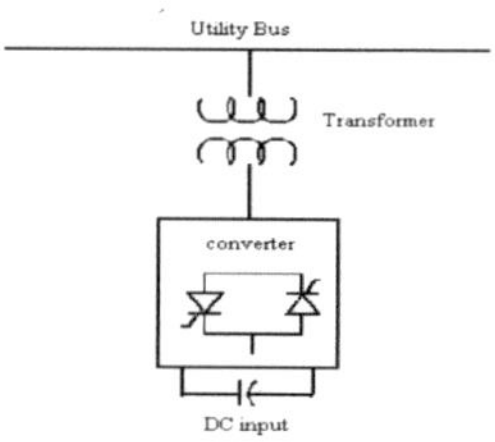

Fig 3 STATCOM connected to Bus

Fig 3 shows a STATCOM connected to a bus as an advanced static VAR compensator. A STATCOM basically consists of a Switching Converter which can both deliver as well as absorb power from the bus, STATCOMs are preferred over other SVCs because they provide a wide range of Reactive Power compensation.It behaves like a capacitor when bus needs power and like an inductor when power is needed to be absorbed from the bus.It eradicates use of bulky passive components and is a very flexible device for improving the voltage profile of the line.

VI. CONTROL SCHEME FOR PV INVERTER

Fig 4shows the block diagram of the control scheme used to achieve the proposed concept. The controller is composed of two proportional–integral (PI) based voltage-regulation loops. One loop regulates the PCC voltage, while the other maintains the dc-bus voltage across SF inverter capacitor at a constant level. The PCC voltage is regulated by providing leading or lagging reactive power during bus voltage drop and rise, respectively. A phase-locked loop (PLL) based control approach is used to maintain synchronization [5] with PCC voltage. A hysteresis current controller is utilized to perform switching of inverter switches.

2

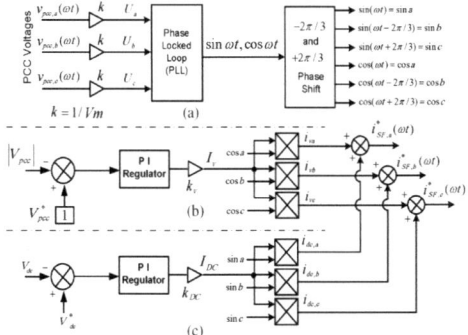

(a)

(b)

(c)

Fig 4 Control Scheme for PV Inverter

VII. SIMULATION RESULTS

To validate the concept presented in the paper, SIMULINK based simulation study is carried out. A Test system consists of integration of both wind energy system and PV Array system.Wind turbines using a doubly-fed induction generator (DFIG) consist of a wound rotor induction generator and an AC/DC/AC IGBT-based PWM converter modelled by voltage sources. The stator winding is connected directly to the 50 Hz grid while the rotor is fed at variable frequency through the AC/DC/AC converter. The DFIG technology allows extracting maximum energy from the wind for low wind speeds by optimizing the turbine speed, while minimizing mechanical stresses on the turbine during gusts of wind.

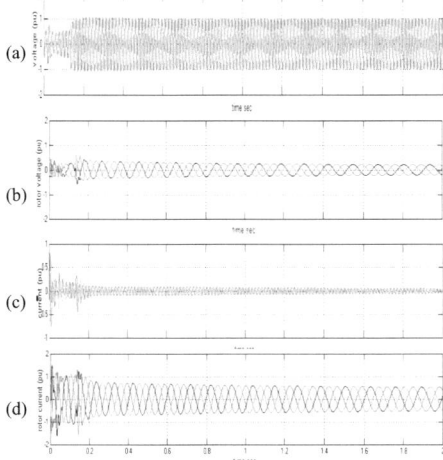

(a)

(b)

(c)

(d)

Fig.5(a) Grid Voltage, (b) Rotor Voltage, (c) Grid Current,
(d) Rotor Current

PV array is made to act as DC capacitor for the STATCOM designed and is interfaced to the system at the 25kv bus as in fig 7. The simulation results are given in Figs 9 to 11.

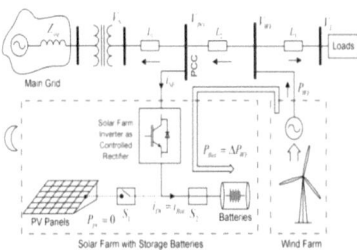

Case1: under normal condition the voltage and current profile in the 25kv bus is as shown in figure 8(a) here after an initial fluctuation for about 0.2 sec the voltages and currents profile are well within the ±5% Pu value criteria. The reason for the fluctuation is due to change over of speed from 8m/s to 14m/s. The simulation is conducted for duration of 2 sec. Also the zoomed view of the voltages and currents is shown in fig 8(b).

Case2: when the system in acting under normal condition single phase short circuit fault is set on the −phase A ‖ line at the 25kvbus at the instant of 0.8 sec. The voltage decreases and current increases after the instant at 0.8 as shown in the fig 10and propagates till the complete cycle of the simulation.

Case3: In this case the fault is implemented at 25kv bus at about 0.8 sec and is allowed to propagate. At 1 sec PV array STATCOM is brought into action and after 1.2 sec system is compensated and system is restored to normal condition after 1.2 sec as shown in fig 11.

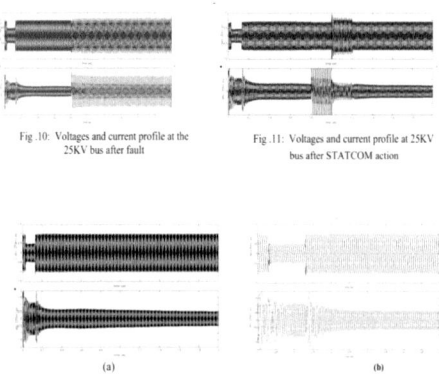

Fig .10: Voltages and current profile at the 25KV bus after fault

Fig .11: Voltages and current profile at 25KV bus after STATCOM action

(a)

(b)

Fig.9: voltage and current at 25kv bus under normal condition

VIII. CONCLUSION

A PV solar and wind plant based distributed generation system with battery storage is studied in this paper. MATLAB/SIMULINK based simulation results confirm the feasibility and effectiveness of the proposed approach to regulate the feeder voltage by exchanging real power through the storage batteries.The proposed strategy of PV SF control will facilitate integration of more wind plants in

3

the system without needing additional voltage-regulating devices. PV Solar farm virtually inactive during night time in terms of active power generation is used to regulate the distribution voltage at PCC within utility specified limits even during wide variations in WF output and loads. This strategy, if implemented, can thus help in generating green power by using two renewable energy sources in tandem.

IX.REFERENCES

[1]Rajiv k. Varma,Vinod Khadkikar, and Ravi Seethapathy, Night Time application of PV solar Farm as STATCOM to Regulate Grid voltage, IEEE Transactions on energy conversion, vol. 24, pp. 983-985, December 2009.

[2] Rajiv k. Varma,Vinod Khadkikar, and Ravi Seethapathy, Grid voltage regulation utilizing storage batteries in PV solar – wind plant based Distributed Generation System, IEEE Electrical Power & Energy Conference, October 2009.

[3] Amirnaser Yazdani Prajna Paramita Dash, A Control Methodology and Characterization of Dynamics for a Photovoltaic (PV) System Interfaced With a Distribution Network, IEEE Transactions on power delivery, vol, 24, No. 3, pp.1538-1551 July 2009.

[4] B. Roberts, Capturing grid power, IEEE Power and Energy Mag.,vol.7, no.4, pp. 32-41, July-Aug. 2009.

[5] R. Fioravanti, V. Khoi, W. Stadlin, "Large-scale solutions," IEEE Power and Energy Mag., vol.7, no.4, pp.48-57, July-Aug. 2009.

[6] L. Wei and G. Joos, Comparison of Energy Storage System Technologies and Configurations in a Wind Farm, in Proc. IEEE Power Electr. Spec. Conf., 17 -21 Jun., 2007, pp. 1280-1285.

4